JESÚS GONZÁLEZ LÓPEZ

PHENOLOGY OF FIG (Ficus carica L.) WITH INDUCED HYDRIIC DEFICITINDUCED

JESÚS GONZÁLEZ LÓPEZ

PHENOLOGY OF FIG (Ficus carica L.) WITH INDUCED HYDRIIC DEFICITINDUCED

BY DRIP IRRIGATION

ScienciaScripts

Imprint
Any brand names and product names mentioned in this book are subject to trademark, brand or patent protection and are trademarks or registered trademarks of their respective holders. The use of brand names, product names, common names, trade names, product descriptions etc. even without a particular marking in this work is in no way to be construed to mean that such names may be regarded as unrestricted in respect of trademark and brand protection legislation and could thus be used by anyone.

Cover image: www.ingimage.com

This book is a translation from the original published under ISBN 978-613-9-46565-1.

Publisher:
Sciencia Scripts
is a trademark of
Dodo Books Indian Ocean Ltd. and OmniScriptum S.R.L publishing group

120 High Road, East Finchley, London, N2 9ED, United Kingdom
Str. Armeneasca 28/1, office 1, Chisinau MD-2012, Republic of Moldova, Europe
Printed at: see last page
ISBN: 978-620-8-30048-7

DEDICATION

Dedicated above all to my parents, Dagoberto González Jiménez and Alicia López Jiménez, who for as long as I can remember have put their lives aside to give me their unconditional support at every stage and obstacle, especially at university. All my achievements will be yours too, I love you infinitely.

To my siblings Lorena, Adriana, Angélica, Emiliano, and my niece María, for always believing in my potential, for giving me your support and love whenever I need it, each and every one of you are the reason for my effort and will to succeed. Without you it would not have been possible to complete my university studies. I love you, thank you.

In memory of my friends Pascual and Francisco, who are resting in peace and were unable to complete their university studies as they would have liked, I remember them with great affection and admiration.

To my friend Israel, for giving me his support, his great friendship all these years and for always having the right words to help me.

To María Fernanda, for all the love, respect and understanding you have given me over the years, you have been a great support in this process.

To my friend Arturo, thank you very much for helping me to carry out several activities of this research, for supporting me academically and personally during these years of university.

ACKNOWLEDGEMENTS

To the Benemérita Universidad Autónoma de Puebla and the Facultad de Ciencias Agrícolas y Pecuarias for having been my second home these years of study and for giving me the opportunity to grow professionally, I take with me great academic and personal experiences.

I thank my thesis director, Dr. Luis Antonio for giving me the tools and support to carry out my research work, I have learned a lot from you academically and personally, thank you.

I thank my advisors, Dr. Carmela, Dr. Sigfrido and M. C Fabiel, for helping me to clear up any doubts I had during my research process and for giving me their time whenever I needed it.

GENERAL INDEX

SUMMARY

The fig tree *(Ficus carica* L) is a crop that, in spite of being very remote and used for a long time as a non-commercial species, has been gradually taking its place in world production. Mexico has the climatic conditions and advantages to become an important producer internationally, but its annual production is low and this could be due, among other things, to the lack of information and experimentation on the influence of irrigation on the production of this species. Worldwide, research has been done on water deficit on yield and fruit quality, but its influence on the phenological development of the crop has been ignored. The objective of this research was to evaluate the phenology of vitroplants and plants per fig stake with water deficit induced by drip irrigation in greenhouse with a 3x2 factorial arrangement resulting in 6 treatments; T1 In vitro 2.3 mm/day, T2 Stake 2.3 mm/day, T3 In vitro 1.7 mm/day, T4 Stake 1.7 mm/day, T5 In vitro 1.1 mm/day, T6 Stake 1.1 mm/day. Stem diameter and height, GDD from pruning to sprouting, third leaf, fifth leaf, fruit emergence and fruit number were evaluated. The variety used was Bellavista, which came from mature plant cuttings and vitroplants obtained in the tissue culture laboratory. They were established in soil in January 2023 under greenhouse conditions. The vitroplants showed greater precocity in the sprouting, third and fifth leaf stages, while the cuttings showed greater vegetative vigour and greater number of fruits.

Key words: *Ficus carica,* phenology, vitroplants, water deficit, earliness.

I. INTRODUCTION

The genus Ficus belongs to the order Urticales, family Moraceae, tribe Ficeae. The fig tree *(Ficus carica* L.) is native to the Mediterranean area of south-west Asia. It is supposed to be the first plant to be domesticated by man (Kisley *et al.,* 2006).

Internationally, there have been several investigations on the effect of water deficit on yield and fruit quality in fig and other crops, thus, in northern China for wheat, savings of up to 25% water have been achieved; In India for peanut crops, they have found that productivity is increased by inducing water deficit at the vegetative stage (20 to 45 days after sowing "DDS"); in Australia in fruit trees, the application of controlled deficit irrigation has produced an increase in water productivity of up to 60%, with improvements in fruit quality and no loss of productivity (FAO, 2011). In the case of fig trees, studies have been carried out under different management scenarios in order to make water use more efficient. For example, the use of plastic mulch (Rodríguez and Valdez, 1999) and high planting densities have been two systems that have fulfilled this objective (De Sousa, 2013; Rivera *et al.,* 2016). Recently Rivera *et al.* (2016) determined fig tree water requirements and crop coefficients (Kc) for the different months of the year in fig orchards established in intensive production systems (2,000 p/ha^{-1}) and drip irrigation in order to schedule irrigation more efficiently. These studies show the importance of deficit irrigation in today's agriculture, however, these techniques must be tested on a regional basis in order to implement efficient water management actions within a given area, and the response to crop yields is highly dependent on climatic conditions.

The Mexican territory has climatic advantages that could place it as an important international producer of figs, above countries that do not have such advantages and place it as a main exporter, however, the production in the country is low and this could be due, among other things, to the little information and experimentation on the influence of irrigation on the production of this species.

Efforts have been made to achieve maximum yields per unit volume of water applied on a site per area basis. This has resulted in water management strategies such as deficit irrigation, where the water supply is less than the crop's water needs and allows for a slight shortage during the stages of development when the crop is less sensitive to water deficiency. Such is the case of figs, which, although it is a crop with low water demands, it is also true that its market and research at national level are little explored.

The world population is growing exponentially, estimated to be 9.7 billion by 2050, a growth of more than 2 billion in about 30 years, which will create a huge demand for food and put a strong pressure on natural resources, so it will be necessary to produce more with the same resources. *In vitro* propagation is used for mass production or storing plant material for later use (Rojas *et al.,* 2004).

In vitro plants propagation is faster, another advantage is that they can be propagated at any date and can be free of pathogens. However, the costs of this method are higher and plants can take more than 2 years to start production, whereas plants propagated 'by stake' can start producing in less than a year (Demiralay *et al,* 1998).

On the other hand, water deficit has an effect not only on yield, but also on phenological stages, vegetative vigour, harvesting times and tropical characters (Fischer and Maurer, 2012).

As mentioned above, for this crop there is not enough information on irrigation experimentation linked to the phenological stages and the way of propagation, as well as the response of the crop to treatments related to the available irrigation lamina in a given region.

II. OBJECTIVES

2.1. General objective

To evaluate the phenology of vitroplants and plants per fig stake with a water deficit induced by drip irrigation and intensive production system.

2.2. Specific objectives

1. To compare the vegetative development of plants propagated by cuttings and vitroplants, subjected to a water deficit induced by drip irrigation and an intensive production system.
2. To determine the accumulation of degree days of development of plants propagated by cuttings and in vitro plants, subjected to a water deficit induced by drip irrigation and an intensive production system.

III. HYPOTHESIS

Variation in water levels during fig growth affects the timing of the onset of phenological stages as well as the length of the crop cycle.

IV. LITERATURE REVIEW

4.1. Origin

Fig *(Ficus carica* L.) cultivation is native to the Asian continent and spread to Mediterranean areas before becoming established in the Americas (Pereira *et al,* 2015).

It is characterised by its versatility to adapt to various climates and soils, as well as being tolerant to drought and salinity, however the best yields occur in dry and hot climates during summer and humid in winter (El-Shazly *et al.,* 2014). Therefore, it could be said that fig is a crop of arid and semi-arid zones.

4.2. Distribution

The main producers are Egypt and Turkey, with 300 thousand and 168 thousand t yr-1, respectively, while Mexico ranks 19th, with a production of 7 thousand t yr-1, on an area of 1340 ha located mainly in the states of Morelos and Baja California Sur, giving a yield of 5.2 t/ha (FAO, 2018).

In the Comarca Lagunera of Durango there is a total of 22 ha of fig trees, with technified production systems; they have technified irrigation, macro-tunnels to avoid cold damage in winter, intensive production plantation (2500 trees) and annual pruning and green pruning to keep the tree canopy compact (SIAP, 2019). ha^{-1}) and annual green pruning to keep the canopy compact (SIAP, 2019).

4.3. Economic importance

4.3.1. Global economic importance

Currently the main producing countries such as Egypt, Turkey and Algeria are in the

Mediterranean and it has become well established in countries such as the USA, Brazil, China, South Africa, Japan and Mexico (FAO, 2018).

In 2018, fig cultivation was recorded in 53 countries, indicating a harvested area of 218,729 ha and an average yield of 6.5 ton/ha (FAOSTAT, 2020).

In 2019, the production obtained from figs in the world was 1,315,588 tonnes, with a harvested area of 289,818 hectares, so the global average yield was 4.5 tonnes per hectare (INTAGRI, 2020).

4.3.2. National economic importance

There are around 1220 ha in Mexico and an estimated production of 6,000 tonnes, valued at 47 million pesos. The largest producers are Morelos, Puebla, Hidalgo and Baja California Sur, with Morelos being the largest producer with 57 % of the national production (Hidroponia, 2015). The 783.5 hectares produced per year in the state of Morelos register yields of up to 5.4 tons per hectare, which originates 3, 713, or more than 31 million pesos per year (SADER, 2019).

4.4. Morphological description

The fig tree *(Ficus carica* L.) is a low growing plant, with a height of 3 to 10 m, it belongs to the family of the Moraceae and to the genus Ficus, in this genus there are more than 500 species, most of them are ornamental plants and some are fruit-bearing species of tropical climates. The fig is a tree with a robust root system, which under the right conditions is quite shallow (20 to 40 cm deep) and at a horizontal distance of about 15 m. The stem is not very elongated. The stem is not very elongated as it tends to be bushy and close to the ground. It has terminal and axillary buds; the leaves are usually large, between 10 and 20 cm, with a thickness of 3 to 7 mm, the syconium envelops the flowers and has an exit orifice named ostil. The true fruits are called achenes, have a hard structure and are small (1 mm). There is a receptacle called achene that envelops the fruit (Lucero, 2018).

4.5. Fig phenology

Phenology could be defined as the study of the relationship between climatic factors and the cycles of living beings (Elias and Castelvi, 2001).

De Cara *et al.* (2007) indicate that phenology is the scientific branch that analyses the periodic biological processes and periods that are related to the climate and the course of the atmosphere at a specific location.

There is no reliable literature on the phenology of figs, comparisons have been made using similar species, managing to classify 5 phenological phases of figs. Phase 1: swelling of buds, phase 2: emergence of first leaves, phase 3: appearance of fruits or sycones, phase 4: fruit ripening, phase 5: beginning of leaf fall (Pucha, 2016).

4.6. Propagation of fig cultivation

4.6.1. Sexual propagation

The fig tree can also be propagated sexually, obtaining botanical seeds to form seedlings, but with this form of propagation the plant fructifies at least 10 years after planting (Mendoza, 2019).

4.6.2. Asexual Propagation

This type of propagation is based on multiplying the plant with plant material from the same plant, which can be cuttings, roots and leaves, among others (Gárate, 2010), which helps to conserve genotypes and have identical and even genetically improved individuals (Ortuño, 2017). The most commonly used forms of this type of reproduction are layering in the fig's aerial zone and cuttings from mature plants.

4.7. Water requirements

The fig tree ranges between 700 and 800 mm per year (Melgarejo, 2000).

It is the root that defines whether or not the plant will be resistant to water stress when it reaches the stage of first fruits, as the availability of water for the root system will determine hollow or fleshy and quality fruits (Tumut, 2002), while an excess of water in this period causes the fruits to crack due to internal pressure (Lobos, 2017).

Fig has a high tolerance to water scarcity, and has even shown good vegetative development in areas of up to 80 mm/year, however in these circumstances it does not produce fruit (De Sousa, 2013; Rivera *et al.,* 2016),

4.8. Water deficit

Occurs when the water demand is greater than the amount used in a given time, or when external and internal factors do not allow water to be available due to water quality, irrigation is scheduled too closely spaced or calculated as deficit for an experimental design or determine the versatility of the soil to retain and crop demand, the soil or irrigation water may be saline limiting water uptake by the crop and may have poor aeration or be flooded preventing water uptake (FAO, 2006).

4.9. General water issues

4.9.1. Field capacity

The amount of water in a saturated soil 48 hours after percolation. Pores larger than o.05 mm help water to drain away, but smaller pore sizes can also participate in the percolation of water into the soil. The term field capacity is valid only in soils with good structure and fast drainage (FAO, 2000).

4.9.2. Permanent wilting point

Refers to the water content of a soil that has lost all its water to the crop and therefore the water remaining in the soil is not available to the crop. Under these conditions, the crop is permanently wilted and cannot revive when placed in a water-saturated environment. On hand contact, the soil feels almost dry or very slightly moist (FAO, 2000).

4.9.3. Evapotranspiration

Evapotranspiration (ET) is the amount of water lost by the crop that must be replaced by irrigation. It is commonly measured or estimated in mm day^{-1} or mm $month^{-1}$ (Allen *et al.,* 1989). Quantification of ETo (potential or reference evapotranspiration) can be done by direct or indirect methods. The most common indirect methods for determining reference evapotranspiration are: Bucket or tank evaporimeter (Pereira *et al.,* 1995).

4.10. Water management in figs

Fig cultivation has quite low water requirements compared to other crops, so using a localised irrigation system is a very viable option for the management and optimisation of water resources. It is suggested to use the fertigation system with which high productivity can be achieved considering that figs require an average of 700 litres of water per year for a good development and quality of the fruit. With this system it is possible to maintain, by means of programmed irrigation, an optimum level of humidity. It is important to have the advice of personnel trained in water management. Furthermore, considering that this species develops well in semi-arid areas, it is necessary to find a way to obtain and optimise water, since in regions with a favourable climate for figs, there are not abundant water resources, and it is necessary to resort to irrigation wells and a whole hydraulic system that can supply the different crops in each region (Flores, 1990).

4.11. Fig irrigation

In terms of water needs, figs are not very demanding, but when the main production locations experience long periods of drought, it is necessary to supply the crop with water through a localised irrigation system (Costa, 2019).

The timing of irrigation is subject to factors such as vegetative vigour, size, soil and rainfall (Flaishman *et al.,* 2002). When fruit ripening begins, it is best to avoid frequent or abundant irrigation as it can lead to rot and poor post-harvest quality (Costa, 2019). Drastically increasing the irrigation rate during ripening will cause the fruit to crack (Melgarejo, 2000). If irrigation is increased in the summer it may cause more vegetative growth and this could affect fruit quality. A saturated soil can lead to large, over-watered fruit, resulting in rotting and wilting (Flaishman *et al.,* 2002).

4.12. Degree days of development

Knowing the period of each phenological stage and the influence of climatic factors is essential to obtain better results (Prabhakar *et al,* 2007). Considering this, temperature is the most important climatic factor in plant phenology. Heat units or degree days of development are the most commonly used variable to determine plant stages (Qadir *et al.,* 2007/Reamar used the term heat units in 1970 and gave rise to different ways to calculate this factor. The concept of heat units, measured in degree development days (GDD), has improved the description and prediction of plant phenological events, compared to other approaches such as time of year or number of days (Cross and Zuber, 1972; McMaster, 1992). The most commonly used method to calculate GDD is the residual method, with the following equation:

$$GDD=((Tmax-T\ min)/2)-Tb$$

Tmax= maximum daily temperature, Tmin= minimum temperature and Tb is the base temperature of the crop. The latter often varies between species and crops. This equation indicates the energy in the form of heat received by the crop in a certain time (McMaster and Wilhelm, 1997). Modifications to the equation have been suggested to enhance the biological significance of the method, such as incorporating a maximum threshold temperature or functions of other environmental factors affecting phenology (McMaster *et al,* 1992).

V. MATERIALS AND METHODS

5.1. Location of the experiment

The research was carried out in the greenhouse of the Facultad de Ciencias Agrícolas y Pecuarias of the Benemérita Universidad Autónoma de Puebla campus Teziutlán (Figure 1). Geographical location at parallels 19° 47' 06" and 19° 58' 12" north latitude and 97° 18' 54" and 97° 23' 18" longitude, at 1607 m above sea level.

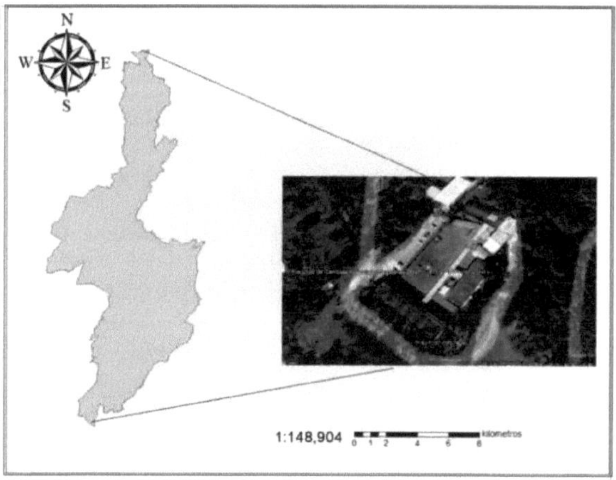

Figure 1. Location of the experiment

5.2. Plant material

The plant material used were plants propagated by cuttings and vitroplants from the tissue culture laboratory of the faculty, the variety used was "Bellavista" due to its economic advantage and its adaptation to intensive production systems. The plants used had been in the nursery for four months and were established in soil in January. They were kept under greenhouse conditions with localised irrigation. As it was an intensive system, the distance between plants was 0.60 m and 1.60 m between rows, with a stocking density of 10, 400 plants- $.ha^{-1}$.

5.3. Setting up and conducting the experiment

The experiment was conducted during the phenological cycle of the crop until the beginning of the reproductive stage, from April to October 2023. In April, the plants were pruned in order to generate and select six productive stems for the new cycle and eliminate the basal or lateral shoots during their development. For the nutrition of the plants, during the establishment of the crop, a background fertilisation was carried out with 100 g of Triple-17, and one kilogram of compost, and then the following formulation was supplied in mg/L of 940 calcium nitrate, 200 potassium nitrate, 340 potassium sulphate, 55 monoammonium phosphate, 490 magnesium sulphate, 15 ferrous sulphate, 4 manganese sulphate, 4.5 borax, 0.4 copper sulphate, 0.4 zinc sulphate, adjusting the pH to 5.6. This nutrient solution was applied by irrigation, always respecting the water deficit in each treatment.

5.4. Experimental Design

The experimental design used was a completely randomised block with factorial arrangement, where the first factor consisted of irrigation rates: 2.3, 1.7 and 1.1 mm per day, and the second was the two types of plants: vitroplants and cuttings. The experimental unit consisted of five plants.

5.5. Treatments

According to the water needs of the crop established by Muñoz *et al.* (2017) fig plants need an average of 0.79 - 1.36 L of water per day for good productivity, however this depends on the climatic factors in the location of the crop.

To evaluate the phenology of the plants and vitroplants of the fig variety "Bellavista" and the effect of the induced water deficit, 2 factors were considered, the first was the type of plant, which has 2 levels (vitroplant and stake), and the irrigation levels factor, whose 3 levels are (100 % of Etc (full irrigation), 75 % of Etc and 50 % of Etc.

Treatment	Factor A (irrigation lamina) mm- day^{-1}	Factor B (type of plant)
1	2.3	*In vitro*
2		Stake
3	1.7	*In vitro*
4		Stake
5	1.1	*In vitro*
6		Stake

Irrigation was provided by a drip irrigation system with emitters with a flow rate of 8 L• and 30 cm spacing between emitters. h^{-1} and 30 cm spacing between emitters.

Irrigation was carried out at an interval of 5 days covering the respective irrigation lamina of each level, which is equivalent to:

Level 1, 100%, 2.3 mm/day= 1:06 hrs irrigation every 5 days

Level 2, 75%, 1.7 mm/day= 58 minutes of irrigation every 5 days

Level 3, 50%, 1.1 mm/day= 29 min irrigation every 5 days

5.6. Variables to be assessed

Number of shoots: The number of shoots emitted by the plant since pruning was counted

Days to sprouting. The days from pruning to the appearance of the first shoot on each plant were counted.

Appearance of third leaf. The date of appearance of the third fully developed leaf was recorded.

Fifth leaf emergence. The date of appearance of the fully developed fifth leaf was recorded.

Total days to third leaf. The days from pruning to the appearance of the third leaf on each plant were counted.

Total days to fifth leaf. The days from pruning to the appearance of the fifth leaf on each plant were counted.

. **Length and diameter of shoots**. Productive stems were measured every 15 days with a tape measure and a digital vernier. Length was measured from the base of each stem to the other end or tip. Diameter was considered below the internode closest to the base.

Appearance of the first syconium per productive stem. The date of appearance of the first syconium on each productive branch was recorded, as well as the number of days elapsed since pruning.

Degrees of Development Day (GDD).). The degree days of development were calculated for the variables of sprouting, days to third leaf, days to fifth leaf, days to first sycon, using the residual method and a base temperature of 13 °C (Wilson and Barnett, 1983).

$$GDD_R = ((TM+Tm)/2)-Tb$$

Where: GDD_R= degree day of development residual method (°C day):

TM= maximum daily temperature (°C):

Tm= daily minimum temperature (°C):

Tb= minimum threshold temperature (°C).

5.7. Statistical analysis

Based on the data obtained, an analysis of variance (ANDEVA) and a Tukey's multiple comparison of means (P<0.05) were performed using the Statistical Analysis System version 9.0 (SAS) package.

VI. RESULTS AND DISCUSSION

Effect of plant type and induced water deficit on fig cultivation.

Table 1 presents the result of statistical analysis for the responses of fig plants to propagation type (Factor A; deficit irrigation levels), and the effect of plant type (Factor B; plant type; vitroplant and stake) for the variables shoot diameter and shoot length.

The table shows that for the variable shoot length in reference to factor A (irrigation level) the first 3 dates of measurement (39, 56, 73 days after pruning) did not show significant differences, while in the following dates (91, 107, 125 days) significant differences were found, and at 141 and 160 days, highly significant differences were found. From this it can be inferred that the effect of the irrigation levels factor had a late onset, however, in the final dates its impact was very significant for the length of fig shoots. As for the diameter and the irrigation lamina factor, there were significant and highly significant differences from 56 days of evaluation, the irrigation levels had a slightly greater effect on the diameter, however, for both variables there was a significant difference in most of the development of the fig plants.

Factor B (plant type) had an effect on shoot length on almost all dates, with date 1 (39 days) being the only one with no significant difference, while dates 2 and 3 had a significant difference and dates 4, 5, 6, 7, and 8 had highly significant differences. As in factor A, factor B, although to a lesser extent, had a late onset for the shoot length variable, but unlike factor A, factor B has more dates with a highly significant difference.

The impact of factor B on shoot diameter was by far the greatest, as there was a highly significant difference on all measurement dates. It can be said that plant type had a highly significant influence on diameter from pruning to the last measurement.

It should be noted that in the interaction of both factors (A* B) there was no significant difference in any of the 8 measurements made for the variables of fruit diameter and length, so it must be assumed that, individually each factor generates significant and highly significant differences in most of the development of the shoots, however the interaction of both has a practically null effect on the diameter and length of the shoots of the fig plants. Rodríguez et al. (2016)

In vitro plants in different ways of production have not had the expected response, so it is essential to experiment by establishing agronomic systems that allow us to obtain more alternative plant material (Rodríguez et al., 2016).

The results of factor B (type of plant; vitroplant and stake) showed great results in terms of the vegetative part, presenting highly significant differences in all the measurements made, so this part of the research can give rise to the fact that there are ways to improve the production of vitroplants, either in the productive part, or, as is the case, the vegetative part.

Table 1. Mean squares and degree of significance for length and length variables diameter on 8 measurement dates for impact of plant type and deficit induced by drip irrigation.

F. V		A	B	A*B	Error
GL		2	1	2	18
L1		15.17NS	8.27NS	2.52NS	19.46
D1	39 days	0.89**	4.38**	0.16NS	0.15
L2		12.84NS	243.46*	8.66NS	59.44
D2	56 days	0.99*	11.99**	0.42NS	0.428
L3		205.68NS	789.59*	79.89NS	126.68
D3	73 days	1.43*	16.55**	0.33NS	0.76
L4		717.15*	1691.92**	285.35NS	199.9
D4	91 days	2.11NS	27.9**	0.77NS	1.20
L5		1521.87*	3067.95**	591.22NS	299.58
D5	107 days	4.06NS	43.65**	1.34NS	1.81
L6		2094.75*	4283.75**	722.85NS	370.52
D6	125 days	10.29*	61.76**	1.98NS	2.52
L7		3436.98**	5430.94**	957.83NS	476.49
D7	141 days	12.95*	73.15**	2.33NS	3.08
L8		5909.96**	6713.41**	1226.98NS	610.15
D8	160 days	17.15*	89.35**	2.84NS	3.88

FV: Sources of variation, GL: degrees of freedom, Variables; D: stem diameter; **L**: stem length, ** highly significant at $P < 0.01$, *significant at $P < 0.05$ and NS: not significant.

Figure 2 shows the average longitudinal shoot growth of fig plants and vitroplants during 8 measurement dates subjected to 6 treatments. After 39, 56 and 73 days from pruning and considering the comparison of means, there was no significant difference from one date to another. From the fourth date (91 days), treatment two (cuttings with 2.3 mm/day^{-1} of irrigation lamina) was the most developed with an average length of 64.3 cm, 47% longer than the vitroplants under the same irrigation lamina (33.72 cm). This trend was maintained during the subsequent measurement dates, reaching a total height of 118 cm. On the other hand, the treatment with the lowest growth was number five, with a final height of 31.46 cm, 73.3 % lower than the best treatment (T2).

It can be clearly observed that the 3 treatments involving vitroplants are the ones that presented the least growth in the measurements made, it can be inferred that, as the lamina decreases, the length of the shoots decreases in the same way, in spite of not being statistically significant, it is observed that T1 had a final length of 56.43 cm, while the T5 treatment was 31.46 cm, a decrease of 44.2 %.

On the other hand, the 3 treatments with plants propagated by stake show a higher growth than the vitroplants in each of the 8 measurement dates. As already mentioned, treatment 2 (stake + 2.3 mm/day) was the fastest growing, with a final length of 118 cm, which translates into an increase of 123.8% in growth compared to treatment 6, which had a length of 47.60 cm.

This behaviour has already been mentioned by *(Hronkova et al., 2003; Nuñez-Ramos et al., 2020)* who write that, in the propagation process, vitro plants can develop abnormally in terms of morphology and physiology, which is not the case *ex vitro*.

Lombardini and Rossi (2019) mention that water stress affects different biochemical and developmental processes in plants, such as decreased photosynthesis (Sperlich et al. 2016), changes in water relations *(Yousfi et al. 2016),* reduced cell division and growth (*Avramova et al. 2016*), as well as the accumulation of sugars, these factors play an important role in reducing productivity. This may confirm what happened in plants subjected to a water deficit, since at irrigation laminae of 1.7 and 1.1 mm, it was observed that the plants had a lower growth which derived from cell division and growth.

Altamirano (2022), when subjecting fig plants of the 'Black Mission' variety to a water deficit, reported shoot lengths between 140 and 180 cm after 150 days of the cultivation cycle, with the lowest values for those that were supplied with 50 % of their water requirement, These results are partially in agreement with those found in this research, since when the irrigation lamina was reduced by 50 %, shoot growth was significantly reduced by up to 40 % in the case of the plants propagated by cuttings, while the vitroplants were more affected with a decrease of 55. 7 % (between treatments 1 and 5). This is in agreement with what is described by Lombardini and Rossi (2019) that the first symptom caused by water deficit is loss of turgor, which causes a reduction in cell size and a reduction in shoot growth.

Figure 2. Shoot length of fig plants and vitroplants with a water deficit induced by drip irrigation.

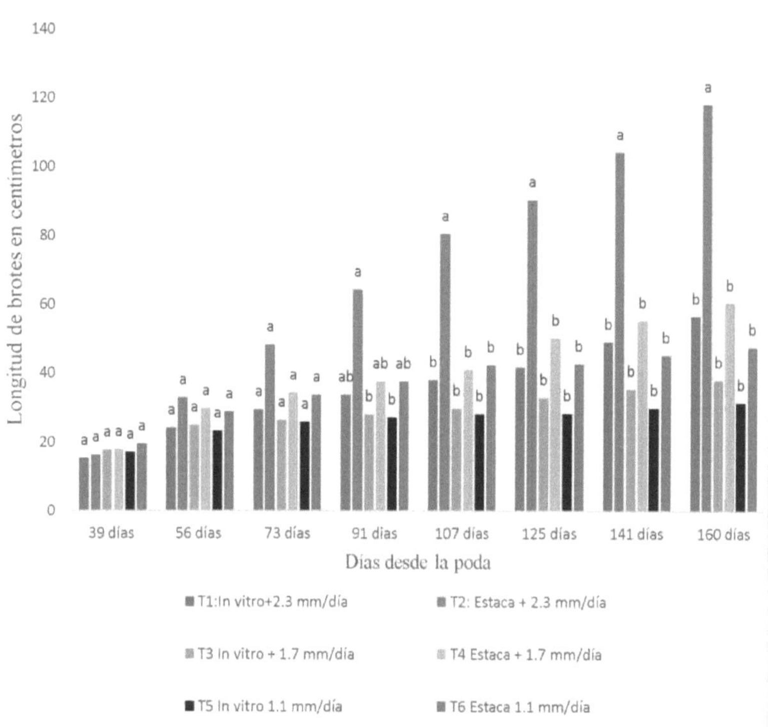

Figure 3 shows the growth of stem diameter of shoots of fig plants and vitroplants during 8 measurement dates subjected to 6 treatments.

The diameters are different from the lengths, since in the first 3 dates significant differences between treatments are observed.

At date 1 (39 days) T6 (stake + 1.1 mm/day) is the treatment with the highest average diameter with 4.6 mm, while the lowest is T1 (vitroplant + 2.3 mm/day) with 2.6 mm.

On the second measurement date, T2 (stake + 2.3 mm/day) had the highest average diameter with 6.1 mm, a 50% increase compared to the previous date (date 1), while the treatment with the lowest growth was T5 (in vitro + 1.1 mm/day) with an average diameter of 4 mm, showing only a 5% increase from date 1. This trend was maintained on all the following dates, with T2 (stake + 2.3 mm/day) being the one with the largest diameter, reaching a final average diameter of 13.8 mm, 40% larger, compared to T2, treatment with vitroplant and the same irrigation lamina.

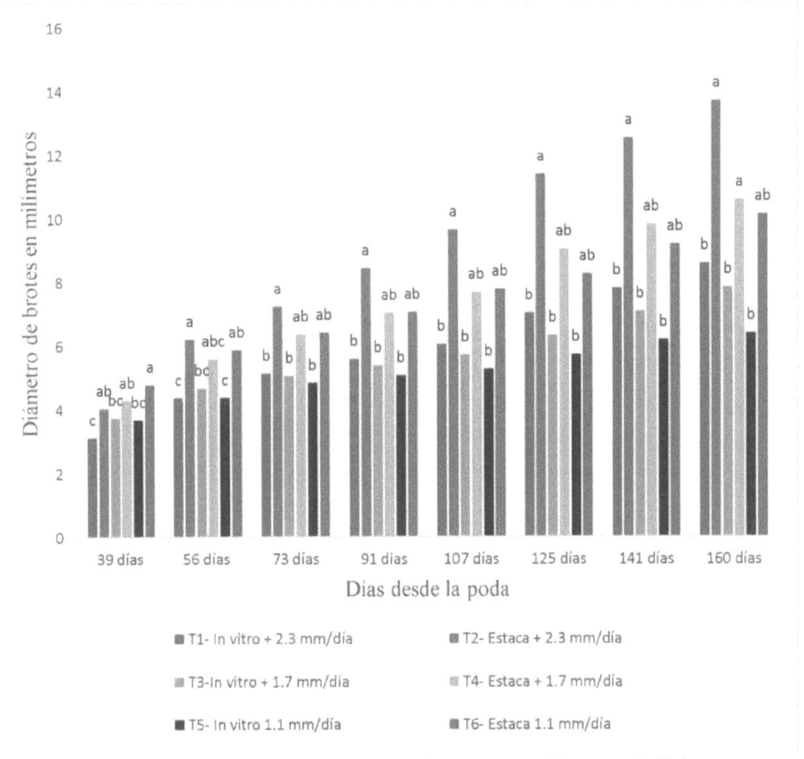

Figure 3. Diameter of shoots of fig plants and vitroplants with water deficit

induced by drip irrigation.

As occurred in the length, the 3 treatments involving stakes were superior to the 3 vitroplant treatments, which could indicate that by reducing the lamina, the diameter of the shoots also decreases, so although there is no significant statistical difference, the T1 had a final diameter of 8.3 mm, while the T5 treatment was 6.2 mm, a decrease of 26%. This is in agreement with (Kulkarni and Pharkle, 2009) who state that water deficit negatively affects leaf area development by restricting the area available for photosynthetic processes, which results in a reduction of aerial biomass. However, it is worth mentioning that, although the treatments with water deficit show a lower vegetative development, this is not the only factor, since the stake-propagated plant treatment with the lowest irrigation level is superior to the vitroplant with the highest irrigation level. From this we can infer that the treatments involving vitroplants were affected to a greater extent by the type of propagation than by the water deficit.

Table 2 shows the analysis of variance for the effect of 3 irrigation levels (Factor A);

2.3 mm, 1.7 mm, 1.1 mm) and plant type (Factor B; in vitro and stake), of the degree days of development for four phenological stages (sprouting, third leaf, fifth leaf, first fruit) and the number of fruits.

The cumulative degree days to sprouting showed highly significant differences for both factors A and B, while the interaction of the two showed no significant difference. The degree days to third leaf showed highly significant differences for both factors. As for the degree days to fifth leaf, factor A shows a highly significant difference.

Regarding the fruit variables, both in the degree days of development for the first fruit (GDF) and in the number of fruits (NF), there was no significant difference for factor A and as for factor B there was a highly significant difference for both cases, the interaction between the factors did not show any significant difference either.

It should be noted that at no stage was there a significant difference for the interaction between factors, indicating that the factors have an effect on degree days of development separately but not together.

Table 2. Mean squares and significance levels for the variables of degree days of development since pruning.

F V	G L	GDB	GD3	GD5	GDF	NF
A	2	6239.64**	2436.16**	7739.83**	456069NS	0.28NS
B	1	8886.57**	11288.77**	13249.77*	17654296**	22.49**
A*B	2	32.45NS	11.4NS	380.52NS	57842.4NS	0.177NS
Error	18	562.90	254.49	949.78	291380.46	0.795
Total	23					
CV	18	12.06	4.52	6.14	23.94	44.53

FV: sources of variation, CV: coefficient of variation, GL: degrees of freedom, ** highly significant at $P < 0.01$, *significant at $P < 0.05$ and NS: not significant, GDB degree days to bud break, GD3 degree days to third leaf, GD5 degree days to fifth leaf, GDF degree days to first fruit, NF number of fruits.

Table 3 shows the multiple comparison of means for the 6 treatments with respect to the variables degree days of development between phenological stages. The *in vitro* plants with the highest irrigation lamina (T1) showed to be the best treatment for the 3 variables GDB, GD3 and GD5, that is to say that T1 is the treatment that generates more precocity in the fulfilment of the phenological stages of the fig, 153, 311 and 451 degree days were accumulated respectively, while treatment 4 was the one that needed a higher accumulation of degree days (214, 377 and 537 respectively). Regarding the beginning of the reproductive cycle, treatment 2 (Stake + 2.3 mm/day) was the one that needed less accumulation of degree days (1272 GD) in comparison with treatment 5 where this stage, in spite of having an accumulation of 3247 GD until the last date of study, the formation of flower buds has not occurred.

When analysing the data obtained for the fig vitroplants, it can be observed that the sprouting of the plants was affected by the irrigation lamina, since when applying 2.3 mm of irrigation this stage needed an accumulation of 153.6 GD, while for deficit irrigation (1.1 mm) I needed 206 GD, which translates into an increase of 17 %. This trend was reflected in the following phenological stages. As for the plants coming from cuttings, their behaviour was similar to that of the vitroplants, with the highest irrigation lamina requiring a lower accumulation of DG, for example, for the appearance of the first fruit, treatment 2 (2.3 mm of irrigation lamina) accumulated 1272.1 DG, while treatment 6 (1.1 mm) required an accumulation of 1612.7 DG, without showing significant differences. 7 GD, without showing significant differences between treatments.

Similarly for the 3 variables of GDB, GD3 and GD5 the plants propagated by stake with the lowest irrigation lamina (T6) were shown to be the latest treatment to reach their phenological stages.

Altamirano (2022) when quantifying the degree hours of growth of the variety 'Black Mission' for two irrigation levels, found that the plants accumulated around 1100 to 1200 degree hours to reach the reproductive stage and found significant differences in those plants that were subjected to a deficit of 50 %, being that the latter were the ones that needed the least accumulation of degree hours to reach this stage. Although in general the variety 'Brown Turkey' evaluated in this research required a greater accumulation of DG, between 1272.1 and 1612.7 for the case of plants obtained by cuttings, the results contrast with those found by this author, since it was the plants subjected to a similar water deficit (50 %) that required a greater accumulation of DG to enter the reproductive stage. Other authors such as Ricardez (2020), when evaluating the duration of the phenological stages of *Capsicum annum* var.

glabriusculum subjected to a water deficit (50 % CC), observed that these had a longer duration than those that grew with the necessary amount of irrigation, which is similar to what we found in figs where the phenological stages evaluated were later in plants subjected to water deficit.

Regarding the variable of number of fruits, both treatments 3 (*in vitro* + 1.7 mm/day) and 5 (*in vitro* 1.1 mm/day) did not present fruits until the studied instance, while treatment 2 (Stake + 2.3 mm/day) presented the highest mean number of fruits with 10 fruits. Rivera-González et al. (2019) when evaluating the yield and its components in figs subjected to water deficit, found differences in the number of fruits per plant, being those that were subjected to deficit irrigation of 75 and 50 % those that presented a greater number of fruits per plant (125, 89 fruits, respectively) in comparison with those that received full irrigation (84 fruits), results that partially agree with those obtained in this work where no statistical differences were found for this variable.

Galindo et al (2018) mention that deficit irrigation maximises yield per unit volume of water applied, but not yield per unit area.

Table 3. Multiple comparisons of means for degree days of development and stages phenological conditions for the type of propagation and irrigation levels.

Treatment	GDB	GD3	GD5	GDF	NF
T1- In vitro + 2.3 mm/day	**153.6 c**	**311.56 d**	**451.14** c	2872.7 a	0.25 a
T2- Stake + 2.3 mm/day	187.63 bc	355.9 abc	485.45 abc	**1272.1 b**	10 a
T3- In vitro + 1.7 mm/day	172.26 bc	332.57 cd	475.78 bc	3215.8 a	**0 a**
T4- Stake + 1.7 mm/day	214.3 ab	377.68 ab	537.46 ab	1305.1 b	**9.5 a**
T5- In vitro 1.1 mm/day	206.58 abc	348 cb	507.4 abc	**3247.4 a**	**0 a**
T6- Stake 1.1 mm/day	**245.83 a**	**388.65 a**	**552.39 a**	1612. 7b	7.75 a
C. V	12.06	4.52	6.14	23.94	44.53
DMS	53.31	35.85	69.25	1213	11.85

DMS honest minimum significant differences, C.V Coefficient of variation, GDB: degree days to sprouting, GD3: degree days to 3rd leaf, GD5: degree days to 5th leaf, GDF: degree days to first fruit, NF: number of fruits. Means with the same letter in the direction of the column are equal according to Tukey's test at $P < 0.05$.

VII. CONCLUSIONS

Propagation type and water deficit induced with 3 irrigation levels affect stem length and the effect of both is individually significant while their interaction is practically null.

The degree days of development are determined by the type of propagation and the level of irrigation applied, being the vitroplants with a higher level of irrigation the ones that presented a greater precocity in the fulfilment of the phenological stages and that also tend to have a lower requirement of degree days of development.

Fruit development and number of fruits is higher for plants propagated by cuttings, while for vitroplants it is almost nil in this production system.

For a higher number of fruits and higher vegetative vigour the best option for the intensive production system is plants propagated by stake with 100% of the irrigation demand.

In order to generate greater precocity in the phenological stages, with less number of degree days of development, it would be ideal to produce with vitroplants and 100% of the water demand of the fig crop.

VIII. LITERATURE CITED

Allen R. G., M. E. Jensen, J. L. Wright, and R. D. Burman. 1989. Operational estimates of reference evapotranspiration. Agronomy Journal 81:650-662.

Altamirano J. G. Eco-physiological responses and irrigation water productivity in fig *(Ficus carica* L) 'Black Mission' under shade netting conditions and controlled saline irrigation by root division. Master's thesis. Master of Science Programme in Plasticulture. Saltillo, Coahuila, Mexico. 65 p.

Avramova V., Sprangers K., Beemster, GTS The maize leaf: another perspective on growth regulation. Trends in Plant Science . 20,787-797 (2015).

Costa A. 2019. The fig tree. Mediterranean fruit tree for warm climates. Madrid, Spain: Ediciones Mundi-Prensa.

Cross H. Z. and M. S. Zuber. 1997. Prediction of flowering dates in maize based on different methods of estimating thermal units. Agronomycal Journal 64: 351-355.

De Cara G. J. A., C. Ruiz L. and A. Mestre B. 2007. Adaptation of the BBCH code to AEMET phenological observation. AME Scientific Journals. 30: 1-7

De Soua A. l.P. 2013. Irrigation Management on the fig tres *(Ficus carica* L.) using the soil water balance. Dissertation (Master Science in Agronomy, Soil Science. Institute fo Agronomy Department Soils. Federal Rural University. Rio de Janeiro

Demiralay A, Yalçin-Mendi Y, Aka-Kaçar Y, Çetiner S. 1998. In vitro propagation of *Ficus carica* L. var. Bursa Siyahi through meristem culture. Acta Hortic 480:165- 167

Elias C. F. and S. F. Castellví. 2001. Agrometeorology. 2ª ed. Mundi-Prensa. Madrid, Spain. 517 p.

El-Shazly S. M. Mustafa N. S. and El-Berry I. M. 2014. Evaluation of some fig cultivars grown under water stress conditions in newly reclaimed soils. Middle-East J. Sci. Res. 21(8):1167-1179.

FAO (Food and Agricultural Organization). 2000. Agriculture: Glossary of soil moisture terms. Economic and Social Department, FAO, Rome. Online: https://www.fao.org/es/#data/QBF. Accessed: 10/09/23.

FAO (Food and Agricultural Organization). 2011. Guidance for policy responsibilities for sustainable intensification of smallholder agricultural production. Online: https://www.fao./es/#data/QCL. Accessed: 07/09/23.

FAO (Food and Agricultural Organization). 2018. Fig production statistics. Online: https://www.fao.org/faostat/es/#data/DBE. Accessed: 10/09/23.

FAO (Food and Agricultural Organization). 2006. Crop evapotranspiration. Guidelines for the determination of crop water requirements. In Fao irrigation and drainage study. https://doi.org/M-56 Accessed 07/09/23

FAOSTAT (Food and Agriculture Organisation of the United Nations). 2020. Fig production in Mexico. Online: https://www.fao.org/faostat/es/#data/QCL. Accessed: 15/09/23.

Fischer R.A. R. Maurer. 2012. Drought resistance in spring wheat cultivars. l: Grain yield response. Aust. J. Agric. Res. 29:897-912.

Flaishman M. A. Rodov V., Stover E. 2002. The fig: botany, horticulture and breeding. Horticultural Reviews-Westport then New York, 34:113.

Flores A. 1990. La higuera. Mundi-Prensa. Madrid. 190 pp.

Galindo, A., Collado-González, J., Griñán, I., Corell, M., Centeno, A., Martín-Palomo, MJ, Girón, IF, Rodríguez, P., Cruz, ZN, Memmi, H. , Carbonell-Barrachina, AA, Hernández, F., Torrecillas, A., Moriana, A., & López-Pérez, D. 2018. Deficit irrigation and emergent fruit crops as a strategy to save water in semi-arid Mediterranean agrosystems. *Agricultural Water Management, 202*,311-324.

Gárate M. 2010. Propagation techniques by cuttings. Thesis in Agronomist Engineering. Ucayali, Peru, National University of Ucayali. 42p.

Hydroponics 2015. Importance of fig cultivation in Mexico. Agricultural articles. Mexico. 4 p.

Hronkova M. H., Zahradnickova M., Simkova P., Simek A., Heydova. 2003. The role of abscisic acid in acclimation of plants cultivated in vitro to ex vitro conditions. Biologia Plantarum 46: 535-541.

INTAGRI (Institute for Technological Innovation in Agriculture). 2020. Fig production in Mexico. Serie Frutales, No. 60. INTAGRI Technical Articles. Mexico. 4 p.

Kisley M. E., Hartmann A., Bar-Yosef O. 2006. Early domesticated figs in the Jordan Valley. Science 312(5778): 1372-1374.

Kulkarni M., S. Pharkle. 2009. Evaluating variability of root size suystem and its constitutive traits in hot pepper *(Capsicum Annum* L.) under water stress. Scientia Hortuculturae 120: 159-166.

Lobos G. 2017. Manejo hídrico en frutales bajo condiciones edafoclimáticas de Limarí y Choapa. inia intihuasi La Serena, Chile, Boletín INIA N° 355.

Lombardini L and Rossi L. 2019. Dryland Ecohydrology. Agronomía Colombiana, 28(1), 71-79.

Lucero G. 2018. Design of an irrigation system using underground diffusers and its effect on ecophysiology and water productivity in fig tree *(Ficus carica* L.) PhD thesis. Centro de Investigaciones del Noroeste, S.C., La Paz, Baja California Sur. 78 p.

McMaster G. S., and W. Wilhelm. W. 1997. Growing degree-days: one equation, two interpretations. Agricultural and Forest Meteorology 87 (4): 291-300.

McMaster G.S., W. Wilhelm W., A. Morgan J. 1992. Simulating winter wheat shoot apex phenology. Journal of Agricultural Science119: 1-12

Melgarejo P. 2000. The cultivation of the fig tree *(Ficus carica* L). Miguel Hernández University of Elche. Ed. IRAGRA, S. A. Madrid. 112 p.

Melgarejo M. P. 1999. El cultivo de la higuera *(Ficus carica* L.), ed. A. Madrid Vicente, Ediciones, Madrid. 116 p.

Melgarejo, P. 1996. La higuera. Author-Publisher. Orihuela. Madrid Vicente, Ediciones, Madrid 83 pp.

Mendoza R.J., L. 2019. Effect of three doses of ROOT-HOR on rooting of fig *(Ficus carica)* cuttings under nursery conditions. Bachelor's thesis. Universidad del Santa, Chibote, Peru. 70 p.

Muñoz V. J. A., M. Palomo R., H. Macías R., M. Rivera G. and G. Esquivel A. 2017. Phenological growth dynamics of fig (*Ficus carica* L.) with high planting densities in macro-tunnels. AGROFAZ 15:133-141.

Nuñez-Ramos J. E., Quiala L., Posada S., Mestanza L., Sarmiento D., Daniels C., Arroyo B., Naranjo K., Vizuete C., Noceda R., Gomez-Kosky. Naranjo K., Vizuete C., Noceda R., Gomez-Kosky. 2020. Morphological and physiological responses of tara *(Caesalpinia spinosa* (Mol.) O. Kuntz) microshoots to ventilation and sucrose treatments. In Vitro Cellular & Developmantal Biology 57:
1 -14.

Ortuño M. 2017. Fig tree varieties. Data for a correct varietal choice. Vida Rural. Year III.

N° 27, 92-95.

Pereira C., M. J. Serradilla, A. Martín, M del C. Villalobos, F. Pérez G., and M. López C. 2015. Agronomic behaviour and quality of six cultivars for fresh consumption. Scientia Horitculturae 185: 121 - 128.

Pereira R., Villa-Nova N., Soares A., Barbieri V. 1995. A model for the class A pan coefficient. Agricultural and Forest Meteorology 76:75- 82.

Prabhakar B. N., S. Halepyati A., B. Desai K. and T. Pujari B. 2007. Growing degree days and photo thermal units accumulation of wheat *(Triticum aestivum* L. and T. *durum Desf.)* genotypes as influenced by dates of sowing. Karnataka Journal of Agricultural Sciences 20(3): 594-595.

Pucha M.L.A 2016. Evaluation of nine accessions of fig *(Ficus carica* L.) in the experimental station of the austro of INIAP, canton Gualaceo province of Azuay-Ecuador. Master's thesis. University of Cuenca, Cuenca, Ecuador. 181 p.

Qadir G., A. Cheema M., F. Hassan, M. Ashraf and M. Wahid A. 2007. Relationship of heat units accumulation and fatty acid composition in sunflower. Pakistan Journal of Agricultural Sciences 44(1): 24-29.

Ricardez L. 2020. Physiological and biochemical effects of water stress on *Capsicum annuum* variety *glabriusculum.* Master of Science thesis. H. Cárdenas. Tabasco. Mexico. 80 p.

Rivera G., Delgado R. G., Macías R., Muñoz V. 2016. Determination of water requirements of the fig crop in drip irrigation and high population in the Lagunera region. AGROFAZ. Coahuila. Mexico. pp.88

Rodriguez J. and Valdez G. 1999. Irrigation of the fig tree. Comunidad Valenciana Spain.

Revista Comunidad Valenciana Agraria. pp:33-38.

Rodríguez R., R. Becquer, Y. Pino, D. López, R. Rodríguez, G. Lorente, R. Izquierdo, and J. González. 2016. Pineapple *(Ananas comosus* (L) Merr) MD-2 fruit production from vitroplants. Cultivos Tropicales 37(1): 40-48.

Rojas G. S., García L. J. and Alarcón R. M. 2004. Propagación Asexual De Plantas. Editorial Produmedios.31 p

SADER (Ministry of Agriculture and Rural Development). 2019. Morelos is the main producer of figs at national level . Enlínea : https://www.gob.mx/agricultura%7Cmorelos/articulos/morelos-principal-productor-de-higo-a-level-nacional#:~:text=Morelos%20is%20the%20main%20producer%20boom%20in%20the%20last%20years. Accessed: 02/08/2023.

SIAP (Servicio de Información Agroalimentaria y Pesquera). 2019. Anuario Estadístico de la Producción Agrícola Online: https://nube.siap.gob.mx/cierreagricola accessed: 5 September 2023.

Sperlich, D., Zhou S., Medlyn B., Sabate S. & Prentice, I. C. (2014). Short-term water stress impacts on stomatal, mesophyll and biochemical limitations to photosynthesis differ consistently among tree species from contrasting climates. Tree Physiology, 34, 1035-1046. doi:10.1093/treephys/tpu072

Tumut S. 2002. Fig growing in NSW. Agfact H3.1.19, first edition, September 2002 Julie Brien, District Horticulturist, Gosford Division of Plant Industries. https://www.dpi.nsw.gov.au/data/assets/pdf_file/0017/119501/fig-growingnsw.pdf accessed 11/09/2023.

Valdés G., Escartín, N., Lorente, M., Malagón, J., and Bartual, J. 2009. Agronomic evaluation and morphological characterisation of selected fig tree material for fig production in Alicante. Actas de Horticultura (54): 135-138.

Villalobos J. A. M., Rodríguez, M. P., Rodríguez, H. M., González, M. R., and Arriaga, G. E.

2015. Phenological growth dynamics of fig (*Ficus carica* L.) with high planting densities in macro-tunnels. Agrofaz: biannual publication of scientific research 15(2): 133-141.

Yousfi N., I. Slama and C. Abdelly. 2016. Phenology, leaf gas exchange, growth, and seed yield in contrasting Medicago truncatula and Medicago laciniata populations during prolonged water deficit and recovery. Botany. 90(2): 79-91.

Printed by Books on Demand GmbH, Norderstedt / Germany